What's for lunch?

Peanuts

Claire Llewellyn

W
FRANKLIN WATTS
LONDON•SYDNEY

Today we are having peanut butter for lunch.
Peanut butter is made from peanuts.
Peanuts contain **protein**, **vitamins**,
minerals and fat.
They give us **energy** and help us to grow
and stay healthy.

What's for lunch?

Peanuts

Please take care as peanuts and peanut products are known to cause an allergic reaction in some people.

This edition 2003

Franklin Watts
96 Leonard Street
London
EC2A 4XD

Franklin Watts Australia
45-51 Huntley Street
Alexandria
NSW 2015

Copyright © 1998 Franklin Watts

Editor: Samantha Armstrong
Series Designer: Kirstie Billingham
Consultant: American Peanut Council
Reading Consultant: Prue Goodwin, Reading and Language
Information Centre, Reading

A CIP catalogue record for this book is available from the British Library
Dewey Decimal Classification Number 634

ISBN: 0 7496 4940 2

Printed in Hong Kong, China

A nut is a hard, dry fruit.

Inside every nut is a **seed**.

Hazelnuts, acorns and chestnuts
are all different kinds of nuts.

Peanuts are not nuts.

They are the seeds of the peanut plant.

The seeds grow in **pods** under the ground.

That's why peanuts are often
called **groundnuts**.

Peanuts grow in warm countries around the world.

In spring, peanut farmers **sow** peanut seeds in their fields.

The seeds are peanuts from last year's **crop**.

The seeds **sprout** quickly and grow into small plants.

The plants grow yellow flowers
on long stalks called **pegs**.
Instead of growing up towards the sun,
the pegs grow down towards the earth.
They push their way through the soil.
Under the ground, the flowers change
into pods and the peanuts start to grow.

After about five months,
the peanuts are ready to **harvest**.
A machine called a **digger**
digs up the plants, turns them over,
and leaves the pods to dry in the sun.

A few days later, a **combine harvester** pulls the pods off the plants and piles them into a wagon.

Peanut farmers save some of the crop
for next year's seed.
The rest is packed into sacks.
The sacks are delivered to factories
all over the world.

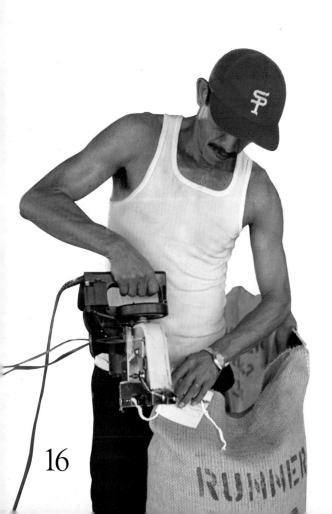

In the factories, the peanuts are checked. Any peanuts that will not be good to eat are removed.

Sometimes peanuts are **roasted** inside their shells.
Or they are taken out of their shells and salted, or coated with honey.

When they are ready to eat
the peanuts are packed into bags.
The bags are sealed to keep the peanuts fresh.
They are now delivered to shops
and supermarkets.

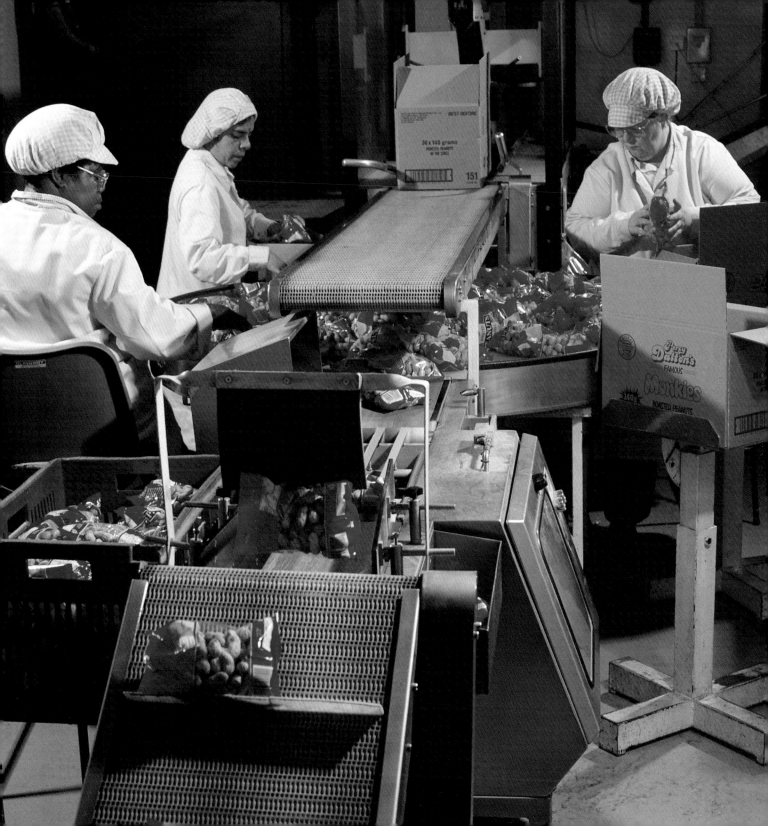

Peanuts can be made into peanut butter. To do this the peanuts are **ground,** and sugar and salt are added to the **paste**. This is mixed until it is very runny and can be poured into jars.

It **sets** into peanut butter.

Oil from peanuts is called groundnut oil
and is used in cooking.
It is especially good for Chinese **stir-fry** cooking
because the oil does not smoke
even when it is very hot.
Peanut oil is also added to hand creams,
lipsticks and paints.

We eat peanuts in all sorts of ways.
Peanuts can be added to biscuits and sweets
or stirred into different dishes
or just eaten on their own.
They are crunchy, tasty and
make a very healthy snack.

29

Glossary

combine harvester a large machine that removes the peanuts from the plants

crop what farmers grow in their fields

digger the machine that pulls up the peanut plants and leaves them on the ground to dry

energy the strength to work and play

ground crushed into small bits

groundnut another name for peanuts

harvest gathering the crop from the fields

mineral a material that is found in rocks and also in our food. Minerals are important for a healthy body

paste what is left when peanuts are ground

peg	the stalk of the peanut plant
pod	the shell that the peanuts grow inside
protein	something found in food, such as peanuts, that helps to keep us healthy
roast	to cook in an oven
seed	the part of the plant from which a new plant grows
set	when something runny becomes solid
sprout	to grow shoots
stir-fry	a special way of cooking food in a Chinese wok or frying pan
vitamin	something found in fruit and vegetables that keeps us healthy

Index

Picture credits: Bruce Coleman 11 (John Murray); Courtesy of Percy Dalton Famous Peanut Company Ltd. (From Cache Collection) 19, 23; F. Duerr and Sons Ltd. 24; FLPA 7 (Silvestris), 8 (L. Wiame/Sunset); Grant Heilman Photography Inc. 6 (Jane Grushow), 10 (Runk/Schoenberger), 12 (Grant Heilman), 14-15 (Arthur C. Smith III); Holt Studios International 8, 13 (Nigel Cattlin); Panos Pictures 16 (Tina Gue), 17 (Jeremy Hartley); Robert Harding Picture Library 26; Zefa 20 (Bramaz); Steve Shott cover; All other photographs Tim Ridley, Wells Street Studios, London.
With thanks to Thomas Ong and Redmond Carney.